The local search series

Editor: Mrs Molly Harrison MBE, FRSA

Trees and Timbers

Fallow deer grazing in the New Forest. The broad-leaved trees are oak and birch; the coniferous trees are young and old spruce. Wood pigeons nest in tree-tops.

Routledge & Kegan Paul
BROADWAY HOUSE, CARTER LANE
LONDON, EC4V 5EL 01-248 4821

The Publishers present their compliments to the Editor and have pleasure in sending for review a copy of the accompanying publication. They would greatly appreciate receiving a marked copy of the journal containing any notice that may be given to the book.

N.B. Reviews should not appear in the press prior to the date of publication.

TREES AND TIMBERS

by

HERBERT L. EDLIN

Local Search Series

Published Price £2.00

Date of Publication 22 November 1973

Trees and Timbers

Herbert L. Edlin BSc (Forestry)

Line drawings by Colin Gibson

London Routledge & Kegan Paul

First published in 1973
by Routledge & Kegan Paul Ltd
Broadway House, 68–74 Carter Lane,
London EC4V 5EL
Printed in Great Britain by Whitefriars Press Ltd, London
and Tonbridge
© Herbert L. Edlin 1973
No part of this book may be reproduced in any form
without permission from the publisher, except for the
quotation of brief passages in criticism
ISBN 0 7100 7590 1

The local search series
Editor: Mrs Molly Harrison MBE, FRSA

Many boys and girls enjoy doing research about special topics and adding drawings, photographs, tape-recordings and other kinds of evidence to the notes they make. We all learn best when we are doing things ourselves.
The books in this series are planned to help in this kind of 'project' work. They give basic information but also encourage the reader to find out other things; they answer some questions but ask many more; they suggest interesting things to do, interesting places to visit, and other books that can help readers to enjoy their finding out and to look more clearly at the world around them.

M.H.

'... all the business of life is to endeavour to find out what you don't know by what you do'

John Whiting *Marching Song*

Contents

		page
	Editor's preface	ix
1	Planning your project	1
2	Getting to know trees	7
3	Many kinds of trees	18
4	Trees in landscape, art and literature	32
5	Forest wild life	41
6	The wood from the trees	50
7	Forests throughout the world	61
	Measuring trees	70
	Naming trees	73
	Handling specimens	74
	Books to read	76
	Acknowledgments	78

For William, Duncan and Sarah

Editor's preface

Woods and forests cover a large part of the surface of the earth and even in our modern cities nobody lives very far from trees. But we take them very much for granted, forgetting how varied they are and how greatly we depend upon them in our everyday life.
This book is to help you to enjoy looking at trees and to understand their importance. It shows you many things you may never have noticed before and — more important — it can help you to find out for yourself. You may perhaps decide to study one particular tree at different seasons, or a forest as a whole, or you may like to work with friends and make an exhibition on all aspects of wood. In any case, you may find that the study of trees and timber can become a fascinating hobby.
Remember how important appearance is in anything we do. However original and interesting your notes or charts may be, however pleasing your drawings, paintings and photographs, they will not really attract anybody unless your work is tidy, your captions clear and your headings well-spaced. Tape recordings, too, need to have titles and other information on a label.
If your project is going to be well-balanced, you will have to use your eyes and your imagination, and you will need to talk to as many different sorts of people as possible. Foresters, timber-merchants, carpenters, gardeners, can all help; people whose windows are becoming shaded by a growing tree have one point of view, those anxious about tree protection and the conservation of the landscape have another. Many people seem to have no point of view at all, they often unthinkingly light fires in woods, drop litter there, and pay no attention to trees. Who knows? Your study might do something to change this.

M.H.

Planning your project 1

How to decide what you want to find out
How to find the facts you need
Putting your work together

Trees, timbers and other forest products, such as paper and rubber, are all familiar objects which are well worth studying. A project about them will test your skill in seeking fresh facts from outdoors, and from everyday life at home, school, or workshop, rather than from books. Everything you discover will have the fascination of being linked to living and growing woodlands.
As a start, read a good general account of raising trees, cutting them down, and using their logs in traditional crafts or modern industries. You can choose some part of this vast field that appeals to you, in which you can yourself discover new facts, not yet recorded.

Deciding what aspects to study
Here are several ways in which you could set about your project:

1 You may decide to find out all you can about the trees in your local park or school grounds, or better still those in a wood nearby.
2 You could study the use of wood and forest products in your home — from the floorboards and roof-beams to furniture, hammer handles, paper of all kinds and even the rubber tyres of your car.
3 You might ask yourself where familiar trees or timbers come from, since many kinds of trees and most of the woods we use are brought from abroad.

Planning your project

Sowing acorns in a nursery bed to raise young oak trees; the acorns will be covered with soil and left to sprout.

One-year-old oak seedling. The upright shoot, springing from the acorn, bears lobed leaves; the roots explore the soil and carry root-hairs.

Planning your project

4   You may look on woodlands mainly as the home of wild animals, birds and rare plants, and link each with a particular tree or forest.
5   Many of the trees you see around were not planted for practical purposes at all. People grow trees because they like them! They have the attractions of all living things. Shaped for a purpose, trees change and grow from season to season. You may enquire how artists have looked at them, or even how photographers make use of their varied forms, with their influence on light and shade.
6   Poets and novelists have expressed people's feelings for the beauty of trees and the mystery of the forest. If your thoughts run that way, there is much to explore in literature.
7   Many countries have their leading or best-loved tree — the oak in England and the pine in Scotland — and in other countries, too, myths and legends are attached to particular trees, such as the well-known Christmas tree, a spruce long treasured in central Europe.
8   Forests also feature in history, as hunting grounds, as the resorts of robbers or rebels, or sources of timber and iron

Transplanting young spruce trees in a forest nursery.

Planning your project

Planting young Douglas firs on felled woodland.

— formerly smelted with wood charcoal — that sustained a nation's wealth and gave it arms or ships for war and exploration.

You cannot work out one simple project which takes in all these viewpoints, so this book asks you to start with the tree itself. Once you have found out how it lives, grows and renews its kind, you can see its place in the wider world of the forest, in timber-trade and industry, art and landscape, and in the history of its homeland.

How to find your information

Growing trees are not easy things to study. They grow taller and larger than humans and can live much longer. Many of them change with the seasons of the year. Most have unusual flowers and their own special seeds. They appear to grow slowly and change little, although measurements prove that many grow fast. Timbers, at first sight, look very alike. You

Planning your project

A horse hauls logs from a clearing in a Scottish sprucewood.

may be puzzled by the way a forester can choose at a glance a piece of ash wood that will make a sound handle for an axe, but reject a piece of beech, knowing that it would snap at the first stroke.

In practice, you must look at several individual trees to learn the full story of only one kind. A seedling shows early growth and form, a well-grown tree gives the pattern of flower and fruit, and felled logs or manufactured objects show the character of the timber. This means that you must travel around to collect your facts — a nursery, a forest and a sawmill or wood-using factory can each supply a fresh store of knowledge. Museums can help, particularly with past uses for wood. Art galleries can show you how painters have appreciated trees, while libraries will, of course, be helpful in many ways.

Recording what you find

Besides making written notes for your report you can learn a lot about trees by sketching, taking photographs and collecting specimens. Trees provide a wealth of material that can easily be gathered and stored. Even in a park where by-laws forbid the plucking of growing shoots, flowers and seeds, you can pick up fallen leaves, flowers or fruits. Along hedgerows and in woodlands you can usually gather all you need, but always remember to leave the vital top-most shoot, or leader of any tree unharmed. Never uproot seedlings except to take one from an over-crowded cluster. Simple methods of pressing or drying, described later (see page 74) make preservation easy. You can also record the appearance of a growing tree by making plaster-casts of twigs, and bark rubbings. You can sometimes get specimens of different timbers and barks from forests, saw-mills, wood-working firms or craftsmen. Each piece will add to your appreciation of timber's natural variations and help build up your story.

Taking photographs of trees, timbers and forests needs care because of scale, light and shade. Unless you have special skill do not aim to make photographs your chief form of record, but try your hand at the simpler subjects, for example, the same broad-leaved tree in summer and winter.

Sketches are especially helpful. You can show the differences in shape between oak and beech leaves, the varied outlines of cones of pine and spruce, or the branching patterns and silhouettes of well-grown trees of different kinds. Draw as much as you can, and add colour wherever it will help.

Be careful to label all your sketches and specimens, and to identify them as closely as possible. Note when and where you found each one, otherwise you may end up with a jumble of pictures and objects that you cannot sort out. Keep a note, too, of the scale or size of everything you draw or photograph.

Always have a purpose in view, but do not reject good material just because it does not, at first, seem to fit. Make a note of it or take a picture, and then put it on one side in case you can work it in later. Keep a record of any book that has proved helpful; you may wish to turn back to it as your theme develops.

A tree or timber project is like a path through a forest — you can never tell just where it is going to lead. Be ready to change course if exciting new facts spring to light on the way.

Getting to know trees 2

Coniferous trees and broad-leaved trees
The age of a tree
The various parts of a tree
Watching the growth of a tree

To understand the life of a tree, you need to follow the progress of one individual tree, from its birth as a seedling until it becomes 'mature' and is felled for timber. But this can easily take a hundred years, and you can only spare a few months for your project. Obviously you cannot follow the life of the same tree directly. And since trees have a yearly cycle of growth you cannot, for example, find fresh flowers in winter or ripe seeds in early summer.
All the same you need to look at actual trees to make your own observations.

Coniferous trees and broad-leaved trees
All common trees fall into two great natural groups:

Coniferous trees have narrow, needle-shaped or scale-shaped leaves, usually *evergreen;* they bear seeds in cones, and yield soft, easily-worked timber. They are also called *softwoods.* Seedlings may have two, three or many seed-leaves.

Broad-leaved trees have broad-bladed leaves that, in most kinds, fall in autumn; their fruits and seeds take various forms, and the timber of most kinds is hard, needing power and skill to work. They are also called *hardwoods.* Each seedling always bears *two* seed-leaves.

Chapter 3 describes many kinds of 'conifers' and 'broad-leaves'. As a starting point, find one young conifer, preferably a pine; you can then look for young seedlings of the same

Getting to know trees

kind and find a log or piece of sawn timber cut from an older tree, to complete your practical observations. Books, particularly well-illustrated ones, will enable you to fill in the rest of this tree's life story and annual pattern of growth.

Choosing a conifer

Scots pine is the commonest conifer and the only native British one that gives commercial timber. Look for it in a wood or on an open heath, and recognise it as an evergreen with needle-shaped leaves set *in pairs*.

These tough, narrow leaves are bluish-green, and the bark of the tree's trunk has a pinkish or orange-red colour. No other tree looks quite like this. Try to find a fairly young pine, between two metres — roughly the height of a man — and eight metres tall. If you live in a town you may be unable to find a pine, so look for a yew or a Lawson cypress in a park, or a Norway spruce, which you already know well as a

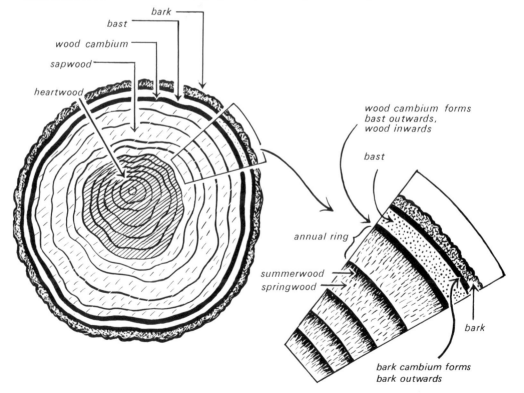

Wood structure revealed by cutting across a log. Left, the main tissues. Heartwood is formed by changes to sapwood. Right, enlarged detail, showing how springwood and summerwood make up each annual ring. In both diagrams, the thickness of the two cambium layers, actually narrow, is stressed to show their positions.

Getting to know trees

Christmas tree. The next chapter, 'Many kinds of trees', shows how to recognise these.
First, make a brief general note of your chosen tree, saying when and where you found it, and how tall it looks. Then sketch it or take a photograph. Next, you need to know how big it is. The section on 'Measuring trees', on page 70, tells you how to do this.

How to find out the age of a tree : rings in timber

Pines grow very regularly, and you will be able to estimate your tree's age by counting the number of branch 'whorls' or groups from the top downwards; each means one year's growth. As a check, find a cut log or stump that is about as thick as the base of your chosen tree. Look at its exposed end for the *annual rings.* Each ring shows as a pair of circles, one pale, one dark. Each pale zone was formed in spring, when the growing tree needed more sap, and is called *springwood.* The dark zone is made later, in summer and farther out to give the tree more support, and this is called *summerwood.* Count these rings to discover how long this comparison log took to grow. Do not be surprised if its 'age' differs by a few years from your previous 'whorl count' estimate, for trees often grow at different rates.
Find out from a botany book how wood carries sap upwards. If you can find a large log, look for dark-coloured *heartwood* in the centre. This was once active *sapwood,* but later served only to support the tree. Both heartwood and sapwood are made up of annual rings.

Other parts of a tree

Now look at the outer bark of your pine tree, and also the bark of the log. What purposes does bark serve? How does its texture alter as the tree gets older? Under the outer bark, but outside the hard true wood, comes a thin layer of pale soft tissue called *bast.* This carries sap *downwards* from leaves to roots, to feed them and enable them to live and grow. Note how thin it is, and think out why the volume of sap that goes *up* through the true wood is far greater than the volume that comes *down* through the bast. Finally, although you cannot see it, remember that an extremely thin cell layer called the *cambium* lies between bast and wood, and forms new rings of wood every year, always on its inner side. You could easily sketch a cross-section of a log, and label all these features.

Roots
Much of your tree is growing out of sight, under the ground. You cannot dig it up to investigate its roots, for it is not your property and such treatment could easily kill it. To find out about your pine's roots you must look at other trees. Perhaps you can find a standing tree that has some roots exposed on the edge of a stream or a cutting made for a new road. You may see a tree that has been blown over by the wind, bringing up the large 'root plate' at its foot, or one that has been uprooted by builders or road-makers. Or look for a smaller tree being transplanted from a nursery.
Look at the exposed roots to see how the larger ones spread out, divide into branches, and finally bear fibrous feeding roots carrying fine root-hairs. See if you can measure how far out they go, and how deep. Exact answers are not easy. The distance outwards may seem great, but the depth, relative to the tree's height, will be surprisingly small. Find out why so many roots are concentrated near the surface. Obviously they help to support the tree, but what else do they do? Make sketches of a tree's root-spread, and add details.

The soil where the roots grow
Now that you are looking below ground, you will want to find out something about the soil. Perhaps you may be allowed to dig a shallow pit. If not, look for an 'exposure' where a cutting has been made vertically into similar ground. Look first at the top layer, just below living grass or dead leaves. This is the *humus* layer, composed of decaying leaf-mould. See how deep it is, what colour, and whether it holds moisture. How does it help to nourish your tree? Below the humus layer you may find sand, clay, gravel, chalk, or possibly broken rock.

Getting to know trees

Note the soil's character — its texture and colours, and see whether it holds up water or lets it drain away. Ask a gardener, farmer or forester if it is a good soil or has some fault — too hard, too sandy, or too wet. Trees are often grown on poor soils because better soils are needed for farming.

Shoot, bud and leaf

Now turn back to the branches of your tree. If it is allowed, break off a portion for closer study; otherwise, get one from

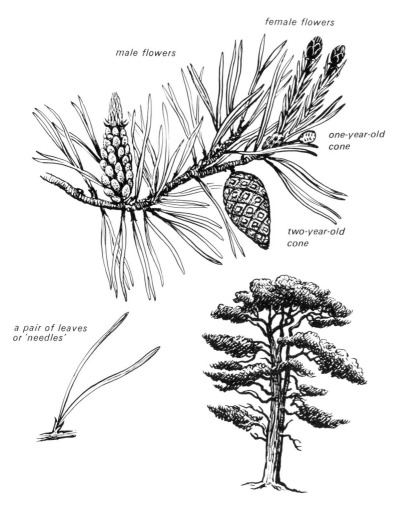

Scots pine. Top, flowering shoot; the female flowers, right, take one year to form small, round, one-year-old cones, and a further year to grow into the large pointed, two-year-old, ripe cones. Below, left: Needles are always in pairs. Below, right: The mature tree has a rugged, rounded crown.

a similar tree where collecting is permitted.

Note how the branch is divided into sections, as the main stem is. Each section has one year's growth, between the joints or 'nodes' where smaller twigs branch off. Can you say how old your branch is?

Now look at the bud at the branch tip. Take it to pieces and see how the tough, brown outer-scales protect soft green tissues within. It is called a 'winter resting bud' and this gives a clue to its purpose. Consider how it is formed slowly in one season to grow quickly in another, extending the shoot by making new cells. The leaves on your pine will all be in pairs, slender, blue-green, and somewhat twisted. Ask your teacher to tell you how leaves use the energy of sunlight to 'fix' carbon dioxide from the air and link it with water to make the carbohydrates — starches and sugars — that the tree needs to build up its substance.

Why do leaves need so much water? How do they send their carbohydrate foodstuffs down to the trunk and roots?

The pine is evergreen, but individual needles do not last for ever. Discover their usual life-span on this branch by counting the years of growth from the tip back to the point where the stem is leafless. The answer varies from one district to another, so do not rely on the figures in a book.

Now make a sketch of part of a branch. You can show the buds and the way the leaves, or 'needles', join the twigs. You will see a little sheath of papery tissue around the base of each leaf-pair.

Flowers — male and female — and cones

Do not be surprised if your chosen branch carries no flowers. Look at other branches or other trees to see what you can find. Trees are large and live a long time. They do not bear flowers and seeds on every branch, or in every year.

You are only likely to find pine flowers, and, indeed, most other kinds open in spring, although withered ones persist through summer. At any other time of year, turn to pictures in a book. Male pine flowers grow in clusters near shoot-tips. Each has many stamens that scatter clouds of yellow pollen.

(opposite)
In this beechwood on the Chiltern Hills, natural seedlings spring up to replace the mature trees as they are felled.

Sketch them, and consider how pollen reaches the female flowers.

Female flowers that receive pollen are quite different. Look for them at the very tip of a newly-expanded shoot. On pine, each is a little crimson globe made up of open scales.

As summer advances fertilised female flowers develop into little round 'immature cones', brown in colour and about as big as a pea. See if you can find one farther down the twig. Still farther down, look for mature cones; these need two years to develop, so before they are ripe the shoot has made two years of growth farther out. Ripening cones are green. As cones mature they turn brown again, in spring. Next, their scales open and they release winged seeds. After this, the cones hang, open and empty, on the tree for a year or so, before falling off. See how many stages you can find in this development from flower to empty cone, and sketch each.

Seeds

Next look for seeds. Newly-ripe cones often remain tight when shut out of doors, but if you dry them indoors their scales will slowly open and you can shake seeds out. Each has a tiny wing attached. If you are lucky you may see seeds sailing down through the trees on a fine spring day. Make a sketch of a winged seed, record its size, and weigh it on a school physics-balance. You may need several seeds to tip the scales! It is surprising how this tiny morsel can grow into a giant tree.

Remember that in one spring a tree may scatter many thousands of seeds. Only a few can ever grow up or the forest would become overcrowded.

Practical work on trees: seedlings and transplants

How does the seed germinate and the tree start to grow? There are three practical ways to learn more:
Grow your own tree.
See how nature does it.
See how nurserymen do it.

Getting to know trees

1 In spring, you can sow freshly-ripened seeds in compost in a plant-pot, cover the seeds with sand, and keep the pot in a cool, well-lit spot until they germinate. After a fortnight, each seed will swell, send forth a root, and slowly rise in the air on a white root-stalk. Then the seed coat will fall off and a tuft of seed-leaves will emerge. Sketch these stages, every day or so. The next group of needles grow *singly*; paired needles do not appear until the next year.
2 At any time of year, you can look for little seedling pines in forests or on waste land near mature, cone-bearing trees. You may be lucky and find all stages and ages, or unlucky and find none at all. This is because squirrels, field voles or rabbits have eaten the lot; birds like pine seeds, too.
3 Somewhere in your district there is probably a nursery where conifers are raised from seed for planting in parks, gardens or forests. See if you can meet the nurseryman and ask him to show you his work. Maybe he does not grow pines, but you can look at other kinds of tree instead; possibly he grows some from cuttings.

Discover how he makes seed beds to raise trees which he calls *seedlings*. When these are a year or two old, he *transplants* them to another bed. This gives each small tree more growing space and obliges it to develop bushy roots. It will need these to get established when it is finally taken from the nursery to be planted in a forest or garden. Can you think how they help?
Note how the nurseryman is careful to keep roots protected and moist. Why is this so important? Nowadays many nurseries sell small trees, usually with roots in containers, at 'garden centres'. These are convenient places to explore and may help to fill in the picture.

Looking ahead
Go back to your own living pine tree, and try to forecast its future. You have seen how pines begin life as seeds and seedlings, are taken from the nursery and grow up in a wood or on a heath. You have studied their timber too. Work out how tall and how stout your tree is likely to become before some future forester describes it as 'mature'. How old will it be,

Getting to know trees

Oak logs arriving at a saw-mill, where they will be sawn into squared timbers like those in the foreground.

what will be the calendar year, and how old will you be then? The answers may surprise you.

Next, look for tree-fellers at work. Note what tools they use, and the order in which they tackle each stage of the job. Follow a log from the stump where it fell to the roadside. How is it moved — by horse, tractor or winch, and how did the men load it onto a lorry for its journey to the saw-mill or the paper-mill? Finally, see if you can find the timber of Scots pine in actual use — in building construction, boxes, telegraph poles, railway sleepers, or 'man-made' wood; chapter 6 (page 50) tells more.

Plantations

Now that you know how a single tree grows, and is eventually cut down and used, you can study how foresters raise trees in plantations. Look for a planted wood, or visit a Forestry

Commission forest on your next outing or holiday.
In some districts — not all — the land is ploughed and drained before the trees are put in. You may be lucky and see the tractor-drawn machines at work. At any time you can measure the depth and distance apart of the furrows and drains and examine the soil that the ploughshare has turned up. Look out for newly planted land, and note what sort of trees are being used (chapter 3 (page 18) will help here). When they were planted, how big were they, and how many years had they spent in the forest nursery? Measure the distance between the trees and check that it is the same both ways. How many square metres has the forester allowed for the growth of each tree, and how many trees does he plant on each standard unit of land area, or *hectare*?
In spring you may come across his planting-gang at work. Note what tools they use and how fast they work. How many trees will each man plant in a day?
Although the little trees look far apart when planted, they soon grow taller and larger and start to crowd one another. After about twenty years' growth, when they are about eight metres tall, the forester will thin them out.
See if you can find a 'thinning' in progress, or at least a plantation that has already been thinned-out, perhaps several times. You can tell that thinning has been done by the stumps, each marking the site of a tree that was removed. Work out what proportion of the trees the forester cut out. Was it 10 per cent or 20 per cent, or even more?
Woodsmen are always busy, but can usually spare time to answer questions if you show you are really interested. Possibly you may meet the head forester and he can tell you his programme of work for the woods you have seen. Make the most of a chance like this and do not forget to take notes. Remember you have two sources of facts for your project — the trees themselves and the people who know and grow them.

3   Many kinds of trees

An oak tree in detail

How to recognise many other trees

Nowadays very many kinds of trees are grown in forests, parks and gardens. Few experts know them *all,* but for a worth-while project you must learn to spot the commoner kinds and to pin-point the special things about each. A good illustrated textbook (see page 76) is essential here. If you are puzzled, as you are sure to be at first, by the scientific names it quotes in Latin, turn to the section entitled 'Naming trees' on page 73.

Study of an oak: its age, size and uses
The commonest broad-leaved tree is oak, and it is the easiest to recognise because of the wavy outline of its leaf. Look at a textbook for other patterns of 'broadleaves' found on *deciduous* trees, that is, those which lose their leaves each winter. Check any leaves you have found against these illustrations.
You cannot easily measure the height of this tall tree, but on a sunny day it will cast a shadow. You will cast a shadow too! Measure your shadow and compare it with your height; let us say it is three-quarters of your height, perhaps 1·2 metres when you stand 1·6 metres tall. Now measure the tree's shadow at the same time. If it is 15 metres long then the tree must be 20 metres tall. Draw a diagram to illustrate how you have worked this out.
Estimating the age of a broad-leaved tree is tricky, since it does not show clear branch whorls like a conifer or grow taller so steadily. Oaks get stouter every year, but on poor soils, outward growth is very slow. See if you can find stumps

Many kinds of trees

or broken branches of other oaks nearby — you must take *oaks* because pines and other trees probably grow faster. Make a count of the exposed annual rings and reckon how long it takes an oak to grow one centimetre thicker (working out from its centre along a radius) in this particular place. Suppose it takes five years. Then if your selected tree has a diameter of 20 centimetres, its radius is 10 centimetres and it is *probably* fifty years old (5 x 10). You cannot make certain without cutting your own tree down; this estimate is near enough.

When you measure the volume of the trunk you may find it tapers unevenly, and it may soon divide into branches. So measure the lower length as though it was a separate 'tree' or cylinder of timber, and then make estimates for one or two stout lengths further up. This is just what the forester and the timber merchant do when an oak is put up for sale. They rarely measure small branches; big logs are worth money, but branches only make firewood.

It is interesting to make a bark-rubbing, too. Hold a sheet of paper against the trunk and rub a soft lead pencil over it. This will quickly pick up the pattern of ridges and hollows, which are difficult to draw or describe in words. Or try taking a photograph on a sunny day when the hollows form shadows, providing, that is, you look at the trunk from the right direction.

When you find an oak log, look for the *rays* that spread out from the centre towards the circumference. All common trees have rays, but in pine and many other kinds they are too fine to see without a microscope. In oak, the rays are thick, and if the log is carefully sawn or split they are exposed on the face of flat planks as shining irregular patches. This is called 'silver grain' and adds greatly to the beauty of oak furniture or panelling. Oak timber has a more elaborate pattern than pine and is used for finer work.

Look for examples of oak in use as good furniture in houses, offices or churches. The structural timbers of really old houses, churches and farm buildings are usually oak. So are the 'ribs' of wooden ships, including Scottish fishing vessels still built today. Your local museum may show oak chairs, tables, chests or cradles. Sherry casks and beer barrels are

Oak catkins and acorns. Male flowers hang down on long catkins. Female flowers, in separate, shorter catkins, ripen by autumn to acorns — the oak's seeds. Pollination is effected by the wind.

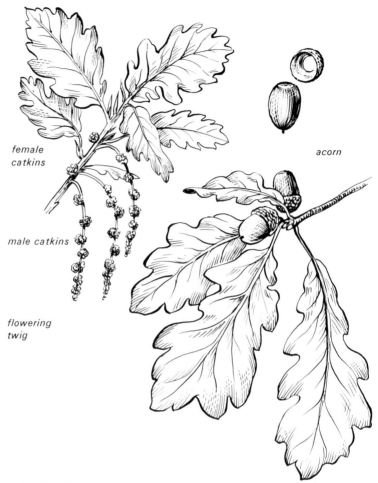

made of oak, too, and so are ladder rungs and wheel spokes. Try to find out why oak is preferred, for these purposes and possibly others. Learn to spot oak by its bold grain, with the silvery patches showing on some surfaces, though never on all of them.

Now go back to your tree to study its shoots. Even in summer you can find the winter buds it is preparing for next year's growth, and in winter they are very clear. Note how they occur at intervals along the stem, usually singly but with a cluster at the tip of each shoot. Clustered buds are a key feature of all oaks and enable foresters to identify them in their leafless winter state. All common broad-leaved trees have distinctive bud-patterns like this. Study a few in a tree identification book.

Many kinds of trees

Oaks usually bear flowers each spring but they do not ripen their seeds every year. Look for flowers in May and acorns in September, but if it is an 'off-year' or the wrong time for your project, find pictures instead.

Oak, like pine, bears separate male and female flower clusters, called *catkins*. Its pollination is carried out by wind. Find the tassel-shaped males. Discover how each female flower, after fertilisation, develops into a one-seeded fruit, called the acorn, a nut sitting on a little woody cup. Compare this pattern of flowering and fruiting with that of another tree — say, a sycamore.

Heavy crops of acorns are only borne once in every four or five years — called 'mast years'. Think of all the birds and beasts that want to eat them, and make a list. See if you can discover what domestic animals like acorns, and say why woods were mentioned so often in William the Conqueror's Domesday Book, written in about A.D. 1086.

You may want to raise your own oak tree. This is quite easy provided you collect a few sound acorns and keep them in a cool, moist place — never in a centrally-heated room — until spring. Then sow your acorns and watch them put forth stout roots and sturdy shoots. In the oak — though in few other trees — there are two seed-leaves that remain in the husk, serving as a food store for the seedling tree. Look for oak seedlings and saplings in a forest or a forest nursery.

Other broad-leaved trees

Every broad-leaved tree has its distinctive group of characters, just as oak does. You will find them set out in any tree-identification book, so only hints are given here. Do not try to study too many trees closely. Select two or three that really interest you and can be found close at hand.

Alder grows only along watersides. Look for 'stalked' buds, roundish leaves, catkins, and seeds ripening in woody cones, which is unusual for a broad-leaved tree. Alder timber was once used for clog-soles, broom-heads and household oddments, because it is soft and easily carved; it was also used to make charcoal for gunpowder.

Ash: Look for hard, black winter buds, set in pairs except for the solitary leading bud at the tip of each shoot. Note the beautiful leaf, made up of many small leaflets, rather like a feather. In autumn, find the single, winged seeds called 'keys'. Ash timber has open-pored annual rings which are very distinct. It is very tough and is used for hammer, pick, axe and spade handles.

Beech is recognised by smooth, metallic grey bark, slender, pointed brown buds, and oval leaves. You can find catkins every spring, but good, sound seed only ripens once every few years. Beech timber has an unmistakable 'trade mark'; its rays show as chocolate-coloured specks on a pinkish-brown ground. Look for it as school desks, chairs, kitchen tables, brush-backs, wooden spoons, platters, tool handles and woodworkers' planes.

Birch is easily known by its white bark and fine, whip-like twigs. Catkins of some kind — male, female or ripening fruit, can be found all round the year. How does it spread, even though it is rarely planted? In northern countries its pale brown, dull-surfaced and hard wood serves for furniture, bowls, platters, tool handles and so on; look out for oddments like broom-heads or cotton-reels. Birch bark has odd uses — for canoes in Canada, and in Switzerland it is used for the huge musical instruments called 'alp-horns', that carry tunes far across the valleys; can you discover others?

Wild cherry looks like garden cherry but grows taller and in woods. Look for white blossom, black juicy fruit, and smooth purple bark carrying horizontal bands of pores used for breathing, known as lenticels. Greenish-brown wood, highly attractive, is used for decorative turnery and fine furniture.

Horse chestnut is grown only for the beauty of its white or pink blossoms; its soft, white wood has no special merit or use. Look for big, brown, paired sticky winter buds, which will open in early spring if you put twigs in a jar of water. Find and sketch the large leaf, 'irregular' flowers, and brown seeds, called conkers, in spiny husks. Horse chestnut comes from the Balkans, where the Turks once fed its nuts to ailing horses.

Sweet chestnut, common only in southern England, has stout twigs, large round nuts, and huge leaves with 'saw tooth'

Many kinds of trees

Flowers and fruits of wild cherry. An insect-pollinated flower, with five sepals, five white petals, many stamens, nectaries, and a central ovary; this ripens to a juicy black fruit holding a hard stone — the seed.

edges. Look for its odd 'male-and-female' catkins in mid-summer and its nutritious nuts. Why did the Romans introduce it? Look for cleft-chestnut fencing, with triangular pales that are still cleft by hand in the Kentish woodlands.

Holly is evergreen and bears thick, waxy leaves to withstand water loss in winter, when the soil is too cold for easy water supply. Note the prickles, on lower foliage only, to stop animals eating it. Look for its flowers in May and work out why only certain holly trees bear red berries. Its hard white wood is used by carvers and wood-turners and it burns well.

Hornbeam grows only in southern England. Watch out for an

irregular trunk, metallic grey streaks down the smooth bark, bent buds and oval leaves. The wood is very hard as its name suggests ('horn', plus 'beam' which is the Anglo-Saxon word for tree). It was therefore used for ox-yokes and cog-wheels in old water-mills and wind-mills. Find out about medieval ploughing and corn-milling.
Nowadays hornbeam is mainly used to make chopping blocks for butchers - why is a hard wood preferred?

Lime is found in parks and avenues. Its leaves are heart-shaped, its twigs often reddish, and its winter buds show only two scales, one large and one small. Its soft, pale brown wood

Wild birchwoods above Loch Arkaig in the Scottish Highlands; note the birch's white bark and airy crown of delicate foliage.

Many kinds of trees

is easily carved and holds its shape well. It is used for piano keys — can you think why? — and for the finest carving. Have you heard of Grinling Gibbons, the seventeenth-century wood carver? Find out about his work, if you can.

Plane is the easiest tree to spot in the squares of London and other cities. Don't mistake it for a sycamore; its lobed leaves and buds are always solitary and are never paired. Look for its catkins and note its dappled bark, which is shed in patches. How does this help it to live in smoky towns?

Poplars include the easily-recognised Lombardy poplar, a very slender tree with small branches, planted for shelter and scenic effect. Look also for the broad-crowned Hybrid poplars planted on rich farmland near rivers to grow fast for making matches, and find out how their logs are cut up.

Rowan or mountain ash is planted in every suburb and grows wild in many woods. Study its large, purplish-brown winter bud, its compound leaf, gay white flowers and scarlet berries. How is its seed scattered? Do you know of any superstitions linked to it, particularly in Scotland?

Sycamore has buds in opposite pairs, broad-lobed leaves, ordinary flowers in clusters (not catkins) and easily-recognised winged seeds that grow in pairs. This is an unusual arrangement. Its pale-brown, smooth-surfaced hard wood is used for furniture, dance floors and the back and sides of every fiddle, although the belly is always spruce wood. It is also called the 'maple' and in Scotland, the 'plane-tree'.

Walnut can be found near farmsteads and is grown for its delicious nuts. Note its compound leaf; crush one to savour its fruity smell and see how the juice stains your fingers. Find its odd male and female catkins and discover how nuts develop. Look for walnut bowls and wood-carvings, and beautiful 'flamy' walnut veneers on the finest furniture.

Willows are of many kinds, but they all show only one exposed scale on winter buds. In spring look for male and female catkins, or 'pussy willows'. Look for the cricket-bat willow, a pyramidal tree with blue-green, slender leaves, grown on good farmland for bat timber. Find out how 'osier' willows are grown for basket-making, and the way their rods are woven.

Various coniferous trees

We looked at Scots pine in chapter 2. What other conifers are you likely to come across? Until you look closely, all conifers seem much alike, and some people use names like 'pine', 'fir', and 'cedar', without a precise meaning. So here is a simple key to nine common groups, using their foliage of needle-shaped leaves:

All these trees, except the larches, are evergreen.

1. Needles in twos, threes or fives: pines.
2. Needles, in clusters, on little knobs (except at shoot tips, where they grow singly):
   - (a) Pale-green, thin, frail, falling in winter : larches.
   - (b) Dark-green, thick, tough, evergreen : Cedar of Lebanon.
3. Needles not seen as separate objects, but as part of a fern-like frond of foliage, touching each other, hiding twigs and buds:
   - (a) Cones round, twigs dark grey: Lawson cypress.
   - (b) Cones slender, twigs red : western red cedar.
4. Needles separate, solitary:
   - (a) Uneven in length : western hemlock.
   - (b) Even in length, set on tiny pegs : spruces.
   - (c) Even in length, no pegs, buds blunt : silver firs.
   - (d) Even in length, no pegs, buds pointed : Douglas fir.

There are many rarities in parks and big gardens; a well illustrated textbook will help you to name them.

Cedars of Lebanon are found in gardens, parks and churchyards. They are usually big, flat-topped trees. Find male and female flowers in September, an odd time for trees to bloom, and collect barrel-shaped cones. Cedar of Lebanon comes from that country and is mentioned in the Bible. Which king bought its strong, fragrant and durable timber to build his temple?

Western red cedar, however, is quite different. Early American settlers in Oregon called it 'cedar' because of fragrant foliage and wood. Look for tiny flowers and slender cones and find its strong durable timber, used for bungalows, greenhouses and roofing 'shingles' or wooden tiles. What colours does it show — when new and when old?

Many kinds of trees

A natural forest of Scots pines, in the Beinn Eighe National Nature Reserve, Ross-shire, Scotland. The snow-tipped peaks are Liathach and Beinn Eighe.

Lawson cypress, as a fern-like evergreen, can be found in nearly every park, and also in many gardens. Note its hidden buds, tiny male and female flowers and round woody cones, no bigger than a pea. It comes from Oregon in the western United States.

Douglas fir is named after David Douglas, a famous Scottish botanist who sent the seed to Britain from British Columbia. Collect its remarkable cone, with a three-pointed leaflet, or bract, below each scale, and look for its strong timber, also called 'Oregon pine', and its broadly-marked plywood; both are imported for building construction and joinery.

Hemlock is a beautiful evergreen from British Columbian forests, with needles of uneven lengths. Its graceful, drooping leading shoot sheds winter snow. It got its odd name because lumbermen thought its crushed foliage smelled like a hemlock plant, a tall white-flowered poisonous weed that grows beside rivers. See what this scent is like, and look for cones. Its pale-brown wood resembles spruce wood and has the same uses.

Many kinds of trees

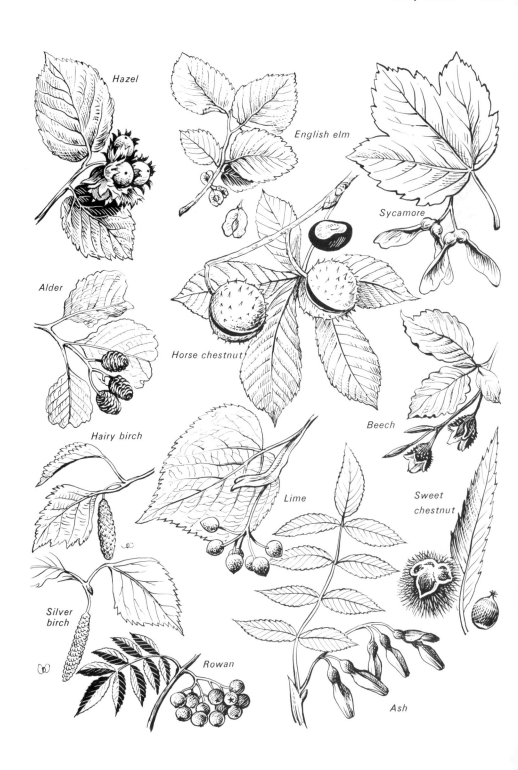

Many kinds of trees

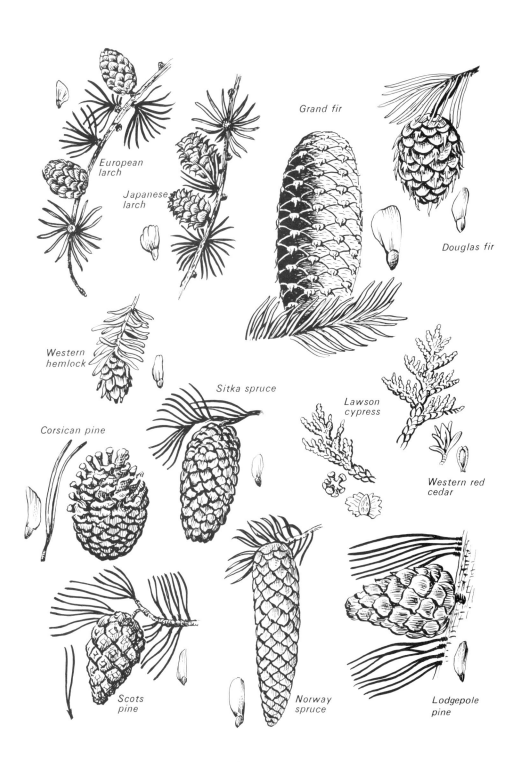

Larches are the only common conifers to shed needles in autumn. All winter through you can tell Japanese larch by bright russet twigs, and European larch from the high Alps, by straw-coloured ones. Watch out for bright, emerald-green young needles in March, yellow male flowers, and white or pink female larch 'roses'. Note its needle colours in summer and autumn, and collect its barrel-shaped cones. Look for its pinkish timber, with darker heartwood, used for strong fences and the planks of fishing boat hulls.

Pines: Scots pine is described in chapter 2 (page 8). Also common are two Black pines, which have needles in pairs and buds that taper suddenly to a sharp point. In forests you can find Corsican pine with long grey-green twisted needles, straight trunk and light branches — preferred for timber. In shelterbelts, which are long, narrow woods grown as wind-breaks to protect houses, crops or livestock, especially near the sea, look for Austrian pine with short, straight needles and many rough branches that help check the winds.

Silver firs are rather scarce trees, but worth a hunt. Check their features — solitary needles, blunt buds and large upright cones seen only in autumn. The beautiful Noble fir, from Oregon, has curled, silvery-blue needles. The Grand fir, from British Columbia, bears long, flat dark-green needles in herring-bone patterns. Both grow fast and yield wood like the spruce tree.

Spruces: All spruce trees have solitary needles on little pegs. If you pull a *living* needle the peg comes away, too; but when a *dead* needle falls naturally it leaves its peg behind, so old leafless twigs feel very rough. You can recognise Norway spruce as the familiar dark-green Christmas tree. Look for long, hanging cones — like the weights on a cuckoo clock — on tall trees. In forests in northern or western England, Scotland, Wales and Ireland the commonest tree planted today is Sitka spruce from Alaska. Look for silvery, blue-green, sharp-pointed needles and small pale cones with crinkly edges. Find a spruce log and work out why everyone wants this rather featureless timber for paper-making, wooden boxes and cardboard for food-packaging.

Many kinds of trees

Tropical timber trees

Very many kinds of wood are imported from tropical jungles in Africa, Asia, and Central and South America. Only an expert can know them all; even he will often have to use his microscope! Three kinds are common and easy to spot:

Mahogany is known by its warm red-brown colour and even texture. Moderately hard, it is easily worked, shaped and curved to give a smooth surface. For 250 years it has been widely used for good furniture of all kinds, and you can easily find mahogany chairs, tables or carvings in your art gallery or museum, and maybe in your own home.

True mahogany comes from a huge tree that bears compound leaves, with about six paired leaflets, and a big woody seed-pod. It grows in the jungles of Central and South America, but nowadays most commercial mahogany is obtained from related trees growing in tropical Africa. See what you can discover about the import of mahogany to Europe.

Rosewood is a beautiful timber showing various shades of brown, with black veins running through it. It draws its name from the fragrance of freshly-cut timber. It is smooth, strong and stable, not changing shape, so craftsmen use it for the best pianos, musical instruments of many kinds, knife handles, brush-backs, and all sorts of decorative woodware. Rosewood trees, which have finely-divided compound leaves and bright flowers like sweet peas, grow in India and Brazil.

Teak, the strongest and most durable tropical timber, grows in India, Burma and Siam. It resists insect attack, fungal decay, and strong chemicals, including acids and alkalis. You may therefore find it used for benches in your science laboratory, also for garden furniture which is exposed to all weather. In Asia, teak is used for house construction, dock and engineering work and ship building. Teak is an even golden-brown in colour, with a coarse grain and an oily surface when freshly cut. The teak tree grows in dense jungles and bears large, heart-shaped leaves and big nuts. Find out how elephants are used to handle it, and how it travels down great rivers, like the Irrawaddy in Burma, to the seaports.

4 Trees in landscape, art and literature

How trees affect the landscape

How artists look at trees

How writers look at trees

Now that you have discovered how trees live and grow, and can tell one common kind from another, you can go on to study why people grow them. Nearly all the trees you see around have been planted — they are cultivated trees, tended by their owners. But here and there, particularly on commons in the south of England, you can find trees and even whole woods growing naturally from self-sown seed, and these too can be fascinating to study.

Making a plan

I suggest you take a definite, restricted area to work over. It could be a public park or even a town square, the grounds of a school or a large country house, a particular farm, the gardens of the road where you live, a set of hedgerows along a field path, or perhaps a small wood. You need to define this area, so you must draw a simple plan. Pace out the ground and measure the angles of the main boundaries — which will often be 'square' or 90 degrees, or nearly so. You may find it simpler to consult an Ordnance Survey plan. Make sure your finished work shows a north point and a scale. Handy scales are 1 : 2,500 roughly equivalent to 25 inches to one mile, to cover several fields, or 1 : 100 (that is, one centimetre on your plan for every metre on the ground) for a garden. Now find out what trees are growing there, and list them by position, kind and size. You may be surprised to find how many small ones you had overlooked before.

Next make a few sketches of the scene, and add, if possible, a few photographs. Besides trees and features such as hedges or

Trees in landscape, art and literature

railings, draw in the outline of any buildings in view.
Next, draw the same scenes in outline, but this time *leave out* the trees. Compare the two versions, and you will realise what a tremendous difference trees make to your surroundings.

Why people plant trees

The people who thought about and spent time and money on planting trees had clear benefits in view. See if you can, at this later stage, read their thoughts. What may their aims have been?

1 Were these trees planted for shelter, and if so, do they shield a house with its garden, a farmstead with its buildings, or the crops and livestock in a farmer's fields? Find

Felling a silver fir with traditional tools. The axe man, standing, cuts away the projecting base and buttresses. Two sawyers then saw through the butt, near ground level. At the centre, a little steel wedge has been driven in, to stop their saw from jamming in the cut.

out about the influence of trees on the flow of the wind. Make your own observations on a windy day. Think of the many effects trees can have. How does a shelterbelt influence the fall and the drift of snow in a blizzard? What about shade for cattle — and for people — on a hot summer's day?

2 Another reason for planting trees, particularly in belts or small woods, is as coverts where pheasants and other game birds can shelter and nest. Possibly there are no pheasants left now, yet the woods remain.

3 They may have been turned into sanctuaries for song birds — I know of one former pheasant covert that has been bought by a Naturalists' Trust. Small woods like this often improve the view very much, especially if they stand, as many do, on hilltops. If you have a tape recorder, try to get some bird songs 'on tape'.

4 Other woods, or even individual trees like hedgerow oaks, elms, poplars and 'cricket bat' willows, may be intended for use as timber. When they are felled the scene will be changed. Do you think the owners will replace them, and if so, why?

5 Obviously, besides having merits as providers of shelter, cover for game, and timber, trees are appreciated for their beauty. Look at your landscape again and try to spot trees or clumps that were selected and placed where they stand for scenic effect. Likely examples may be a slender Lombardy poplar, a pyramidal spruce, or a copper beech with lustrous purple-brown summer foliage.

6 Trees and small woods that are important in the landscape may be protected by the law. See what you can discover about Tree Preservation Orders made by local councils. Perhaps some of the trees you study are protected in this way. Why does the local authority take this trouble, and how can it act when the trees must eventually be cut down?

In your sketches you will already have shown the main differences in tree form. The Lombardy poplar will make a tall thin column, the spruce a tapering cone, and the beech a great irregular, upturned bowl of leaves standing on a stout, erect trunk. The details may vary, according to the season of

Trees in landscape, art and literature

Hedgerow trees and small broad-leaved woodlands give English landscapes their unique character, and provide shelter for farm crops and stock. A view from the Long Mynd in Shropshire.

the year. Evergreens look much the same all the year round. Deciduous trees show frameworks of boughs, and the tracery of finer twigs very clearly when they are leafless in winter; in summer, only dense masses of foliage strike the eye. Notice the colour changes in our common broad-leaved trees: pale-green in spring, they become darker in summer and turn yellow, brown or orange as their leaves fade in autumn. By making paintings or using colour film in your camera, you can record these swift changes, provided you are ready for them at the right time of year.

How artists record trees and use tree shapes

How have the world's great landscape artists drawn and painted trees? Look in your local library, visit an art gallery and collect coloured postcards as guides to what they have done. Trees present real technical problems to skilled painters.

'The Timber Waggon' by Paul Sandby (1725-1809). An eighteenth-century artist's impression of an ancient forest of oaks and beeches, with oxen drawing a high-wheeled timber cart, probably near Windsor.

Texture is difficult, for you cannot draw every leaf or twig. Colour sets problems, even on the same day, for it varies with sunlight and cloud shadow. No two shades of green are alike, and a tree's outer foliage shades its inner leaves, as well as the ground below. Again, a tree can look quite different when it is swept by the wind from how it looks on a still calm day. Great painters know all this, and make their trees come alive, reflecting the moment they were seen, the weather of that day, and the countryside in which they grew.

The greatest English painter of trees was John Constable, the son of a Suffolk miller, who lived from 1776 to 1837. You can easily find reproductions of his works, or possibly see originals in the National Gallery in London and other collections. Note how he caught the play of sunshine on the great oaks and elms that grow beside East Anglian rivers. In

Trees in landscape, art and literature

the same picture one tree will be brightly lit, another made dark by cloud shadow, while a third is seen outlined as a silhouette against a clear white sky.

In France, the Impressionist school of painters in the late nineteenth century saw trees with a different eye. They did not attempt to draw them so precisely. Jean Baptiste Camille Corot (1796-1875) gave the foliage of birches an airy grace, with irregular outlines blending into a background of sky. Alfred Sisley (1839-99) used more 'solid' poplars with great effect to give structure to riverside scenes along the Seine near Paris, which would otherwise have appeared flat and dull. Van Gogh (1853-90) loved to show trees bathed in brilliant sunlight casting dark shadows. In Britain, in our times, Rowland Hilder has favoured tall beeches, in their leafless winter state to give contrast of form to the rounded downs and regular fields and farmsteads of the Sussex Weald.

'The Mill at Charenton' by François Boucher (1702-70). A French artist's impression of a fantastic timber-built corn mill, set amid quaint forest trees that have been repeatedly lopped to yield firewood.

Look for examples of these works, and consider how each painter has his favoured kind of tree, as well as his chosen territory and techniques. Often trees are used to give point to a broad landscape of which they form only a small part. A well-known example by Cézanne (1839-1906) is 'Mont Sainte Victoire', in which the outlines of isolated pines emphasise the steep slopes of a distant mountain.

You might like to paint a favourite tree yourself, on a sunny day or a misty one, and at different seasons of the year. Try looking at it from above, out of a window perhaps, and notice the different effect when you lie down under it and look up into the branches.

Designers often use tree forms as decorations. Leaf shapes, patterns and colours can be found on wallpapers, pottery, trays, carpets, curtains and dress materials. Solid forms, such as acorns and pine cones, may be repeated in carved woodwork or even in cast iron or sculptured stone. An old cathedral, built and furnished by country craftsmen, is a good hunting ground for tree representations. You can find them, for example, on the undersides of *'Misericords'*, those quaint fold-away little seats at the back of medieval choir stalls.

Trees in literature, poetry and music

> Once upon a time, on the edge of a vast and mysterious forest, there lived two old peasants who. . .

You will recognise this as the opening line of many folk-tales, although sometimes the principal character is a witch, a king like the famous Wenceslas, an outlaw such as Robin Hood, or even a fairy prince. Legends, myths, and ballads may all start in this same way. Somewhere in the background of most early European literature lurks this wolf-haunted primeval forest, which our ancestors had to tame and fell before they could live settled lives on land hard-won from the trees. American literature, too, has a similar feeling for the vast trackless forests that faced the early settlers in Canada and New England, aptly expressed in the lines from the poem *Evangeline* by Henry Wadsworth Longfellow (1807-82):

> This is the forest primeval. The
> Murmuring pines and the hemlocks.

When folk-tales and legends developed into modern novels or plays many writers — although by no means all — continued forest traditions. Thomas Hardy (1840-1928) in his novels *The Woodlanders* and *Tess of the D'Urbervilles*, wrote from intimate knowledge of living trees and men and women who, in Victorian times, lived and worked amidst them. He knew hurdle-makers, tree-fellers, and people who stripped bark off fresh-felled logs, to change hide into leather at the tanneries. If you visit his birthplace and early home, in the shadow of a great forest east of Dorchester in Dorset, you will see how easily trees came into his thoughts and plots.
In many of Shakespeare's plays, too, there is a sense of familiarity with the forest, particularly in the comedy *As You Like It*. Remember that the author lived at Stratford-on-Avon beside the Forest of Arden, and once got into trouble for poaching deer in Charlecote Park nearby. See what other novelists, poets or playwrights show this forest sense. Turning to America again, can you imagine *The Last of the Mohicans* by Fenimore Cooper (1789-1851) who lived close to the Indian frontier, being thought out without its woodland surroundings?
In descriptive, narrative and romantic poetry trees are usually described as scenery, part of the background to the action or landscape that forms the main theme. So, in *The Birch Grove*, written by the Welsh poet Dafydd ap Gwilym in the fourteenth century, translated in modern times by Grace Rhys (1865-1929), we read from one verse of the poem:

> Ah, the pleasant grove of birches,
> A pleasant place to tarry all the day;
> Swift green path to holiness;
> Place of leaves on branches deftly strung —
> Tapestry meet for proudest princess;
> Place of the thrush's voice, king of song.
> Place of the fairest-breasted hill,
>   green place of tree-tops,

> Place set apart for two, free from jealous strife;
> Veil that hides the maiden at the wooing,
> Full of delight is then the green birch grove.

In lyric poetry, the writer either aims to put into words the character of the tree, or to express his own feelings towards it, as in this example from *The Lost Bower* by Elizabeth Barrett Browning (1806-61):

> Here a linden-tree, stood, bright'ning
> All adown its silver rind;
> For as some trees draw the lightning,
> So this tree unto my mind
> Drew to earth the blessed sunshine from the sky
> Where it was shrined.

In your own reading you will have found many expressions of feeling for trees. See what your favourite poets have written about them, and quote suitable verses in your project. Then try to compose a short poem of your own, featuring a tree.

Finally we have the links between trees and music. For example, Beethoven's Pastoral Symphony, composed in 1808, echoes spring breezes stirring cool green woodlands beside peaceful fields. Think, too, of woodland birdsong in spring and the rustle of faded leaves in autumn.

Can you think of any songs, carols or ballads that mention trees? One good example is a north country ballad with a haunting refrain, that captures the home-sickness of an exile away in a southern city:

> O! the oak and the ash,
>   and the bonnie ivy tree
> They flourish at home
>   in my ain countrie.

Forest wild life   5

Plants living in a wood

Some birds living there and others visiting

Animals in a wood

Wherever trees grow together, they provide a home for plants, birds and beasts that you rarely see elsewhere, or at least a nesting place for creatures that seek their food in more open country. In order to study nature in the woods, you must be free to visit them. This is not always possible, because many privately-owned woods are preserved as game coverts, where pheasants breed and feed. Fortunately, in most parts of the country there are public woods that can be explored at will. There are a few exceptions, such as certain nature reserves.

Choosing a wood to study in detail
First find an 'open' wood. It may be part of a common or large park owned by a local authority, a National Trust property, or one of the national forests through which the Forestry Commission has opened *Forest Trails*. It need not be a large wood. A small area, under trees, if varied, will hold more than enough for your study. You may find one near at hand that you can visit often, or you may only be able to reach your wood after a long journey from home, or see it only on holiday. Plan your time accordingly.
Choose, if you can, a place that is seldom disturbed by people. This may seem difficult, but if you examine paths and watch passers-by, you will find that nearly all visitors keep to beaten tracks. A few yards away you will find rare plants thriving on untrodden ground, and possibly birds nesting, for they too know where they will not be disturbed.

Next, you need a few simple, well-illustrated guides to help you identify wild plants and animal life. Do not aim at an advanced study, unless it has a special appeal to you. There is so much to search for that you will only have time to find a plant or animal, describe it and ask a few simple questions about its life history. Make a brief general description of the wood you have chosen. Identify the main trees and estimate their likely age, by methods described in earlier chapters.

As you go into the wood you will notice how the intensity of daylight lessens. This is very important if you take photographs. Remember that you will need a much longer exposure under the trees than out in the open — twice as long at least.

This lack of bright light is very important to plants also, for they must have light for energy and growth. All true woodland plants are adapted in some way to live without direct summer sunlight. Find out how the different kinds manage this in their various ways.

Spring plants, climbing plants and 'lower' plants
If you start your project in spring you are sure to be attracted by a group of gay flowering plants. They include white windflowers or wood anemones, yellow primroses, bluebells and pink campion. See how many of these you can recognise and make coloured sketches.

Because these spring-flowering plants are active for only short sunny spells, while the taller trees are leafless, they have to store food all through summer and winter too. The bluebell does so by forming a fleshy white bulb surprisingly deep down in the soil. Look closely at the other plants to see what storage organs each one develops to help it survive. Other woodland plants do not open their leaves or flowers until later. Most of these have pretty, fern-like foliage with much-divided leaves that catch every ray of light that filters through the upper canopy of tree leaves during summer. Look for one or two of these 'leaf-mosaic' plants, name them, and sketch the leaf patterns of a small group to show how the leaves intermesh, without overshading one another. Examples are the pink-flowered herb Robert and the white hedge parsley.

Forest wild life

Climbing and scrambling plants are peculiar to woods and their fringes, because they must have the support of trees and bushes. See how many you can find and note their different climbing methods. Ivy bears little rootlets all along its stem that anchor it to the rough bark of the tree that supports it. See if you can find its very different 'mature' form, bearing oval leaves, green flowers and black berries, that only appears when it has climbed into the top of a tree. Honeysuckle climbs by twining its woody stem around a tree. Wild clematis or 'traveller's joy' holds itself up by turning its leaf

A woodpecker feeds a nestling at the mouth of its nest-hole. This bird's strong claws and stiff tail support it on tree trunks, where it taps decaying wood, using its powerful bill to secure insects. The nest hole is pecked out inside a tree trunk, to give security to eggs and youngsters.

stalks into tendrils; later, the leaves fall off but by then its woody stem has grown over forks where branches begin, so it never falls. Look for climbers using other methods — prickles, rough stems and so on.

Woodlands are particulary rich in what botanists call 'lower' plants, because they have simpler structures than 'higher' seed-bearing plants and trees. These include ferns, fungi, mosses, algae and lichens. In place of flowers and seeds they bear curious structures called sporophores which shed spores and so enable them to spread. With the help of textbooks you can learn about their life histories and methods of nutrition.

Look in summer for the broad fronds of ferns, which open in moist shady woods and bear their sporophores as brown, round or oblong structures on their undersides. In autumn find toadstools, the sporophores of various fungi that usually feed on dead wood and cause its decay. But some grow on living trees, causing disease and death.

On the forest floor, where there is too little light for larger plants, and on rocks where there is too little soil, you may find a dense turf of mosses. In moist places, like streamsides, you can spot 'liverworts,' with their more leafy fronds. The tree trunks themselves may carry green algae or grey lichens, which are remarkable dual-plants, being combinations of an alga with a fungus.

Birds that live in the wood and others that visit
In the woods you will naturally seek birds that are specially adapted to life amid trees. What you find will depend a great deal on time of year. Those that nest in tree-tops are mostly 'residents' that stay in Britain the whole year round. You will only see them at their nests in spring and early summer, when they are laying eggs and rearing young. But they roost in trees at all seasons, even though they usually feed on the ground, often out in the fields. Most are big conspicuous birds, which rely for their safety on swift and often noisy flight, so they are fairly easy to spot — although not always so easy to name.

Look for the rook, an all-black bird which lives in flocks, and

Forest wild life

A buzzard brings its dead prey, a young rabbit, to its hungry nestlings in their tree-top nest.

hunts for grubs and worms in farmers' fields. At dusk, rooks fly home to their home woods. In spring they build nests of twigs, always in colonies called rookeries, where they make a great din by continuous cawing. See what other members of the crow tribe you can find, and discover how their habits differ. Jackdaws, for example, will only nest in *hollow* trees or on buildings or cliff faces — they need a hole. Then there are jays, magpies, carrion crows and possibly hooded crows, although the last-named only live in the west of Scotland, Ireland and the Isle of Man.

Wood pigeons are also common tree-top nesters, but they hide their nests close to tree trunks. Besides the common kind, or ring-dove, you may find the stock dove — which

needs a nesting hole like the jackdaw — the collared dove that immigrated from eastern Europe in the 1960s and the pretty turtle dove that comes as a summer migrant from Africa. Birds of prey are rarer and much harder to find. Look for *owls* — of several sorts, kestrels and sparrowhawks.
Lower down among tree trunks lives a fascinating group of insect-eating birds which get food from wood and bark. Best known are woodpeckers, which cling to trees with their very strong claws — two pointed forwards, two backwards — supported also by a stiff, stout tail. With their powerful neck muscles they drive their sharp beaks into rotten wood, when they hear grubs boring, and then pull these grubs out with their long forked tongues. Tree creepers crawl up tree trunks seeking insects hiding in bark crevices; then they fly down to the foot of the next tree and start the same performance again. Nuthatches can crawl up, down or sideways on tree trunks to find grubs, while tits seek insects and small seeds far out on thin branches.
The woodland undergrowth of shrubs, low branches and tall plants has its own group of year-round residents, including thrushes, blackbirds, robins, wrens and hedge-sparrows. In spring these are joined by many migrants from tropical Africa who come to nest and feed their young on the increasing insect life. Look out for warblers and the sweet-voiced nightingale.
Down at the ground level you may find the nests of large sporting birds, particularly pheasants which are preserved on many estates to provide autumn shooting. Also the woodcock, a brown bird with a long straight bill that, when disturbed, will sometimes carry its young up into the air, held between its breast and its toes — I've seen it done!

Beasts
Woods hold many four-footed furry creatures, but most are much shyer than birds, and all you may see are signs of their presence, rather than the animals themselves. Learn how to spot and interpret these signs and to link them with different creatures. In winter snow, or on soft mud at any time of year, you will find spoors or foot-prints. Fox, badger, rabbit and

Forest wild life

stoat all make distinctive imprints, and in some woods you can find the sharp hoof-prints of deer of various kinds. Beasts that dig, like badgers seeking grubs or rabbits making burrows, leave clear scratch-marks in soft ground, and you may even chance upon a fox's earth or badger's sett. Droppings, too, reveal the presence of mammals by their size and shape, from the round brown pellets left by rabbits to the irregular black ovals formed by members of the weasel tribe. If you are lucky and move quietly, you may see a fox on the prowl, at dusk, or glimpse a herd of deer emerging from the trees to graze on an open ride.

Squirrels are easy to watch, for they are apt to scamper up

The agile red squirrel is marvellously adapted to range the tree-tops, where it eats pine seeds and nuts.

The agile pine marten can catch squirrels in the tree-tops.

trees from their feeding places at ground level. You can follow their amazing acrobatic progress as they leap from one thin branch to another. See if you can spot their dreys, or tree-top nests built of twigs and leaves. There they shelter in winter and raise their young in spring and again in early summer.

In Scotland or East Anglia you can still find native red squirrels; elsewhere most belong to the grey species, introduced in about 1880 from North America. Work out how squirrels store nuts — but often fail to find them again! How does this help trees to spread?

Forest wild life

Find out from a forester which animals he seeks to control. Does he fence rabbits out of his plantations, and are deer so numerous that some must be shot each year to lessen the damage to trees or farm crops? A gamekeeper has different duties and another point of view. If he is raising pheasants he will not want to see foxes prowling around, although the forester would not worry. Why not?

Follow up the life stories of woodland beasts in illustrated books, and visit a local museum, or better still a wild-life collection in a zoo.

A herd of park deer, although tame, will give you a good idea of what their wild cousins look like, how they feed, run and breed.

There are many other creatures that you *might* study in the woods. These include reptiles like snakes and lizards, amphibians such as frogs, bats that shelter in hollow trees, moles below ground, and a whole host of insects, snails and lesser creatures. One of these groups may be your favourite field of study; if so the forest is a good place to look further.

6   The wood from the trees

Sawing, smoothing, carving, turning and veneering
Seasoning and preserving wood
Various kinds of 'man-made' wood
Other materials we get from trees

Wood in one form or another comes into our activities every day. Houses have wooden floors and rafters to support their roofs. We sit on wooden chairs and use tools with wood handles. Newspapers, wrapping papers, books and writing paper are all made from wood pulp, obtained from tree trunks.

Felling trees and changing them into useful timber
The best way to learn how people change trees into timber and other wood products is to look first at simple hand tools. In large-scale practice these have been replaced by fast-moving machines whose actions are hard to follow. Take a look at axes, saws, planes, and other tools in your school work-shop or at home. Probably you know how to use most of them already. If not, see if you can get some practice, or at least watch an expert. Only then will you really understand what each hand tool does and be able to follow the work of a machine that does the same job. Until recent times the axe was a major felling tool. See what you can discover about the materials used for making axe-heads in different periods. Many museums hold Stone Age and Bronze Age axe-heads, but ancient iron ones are rare because iron rusts away in damp earth. Look at modern axes with steel heads. Note their design and study the strokes that tree-fellers use.

(opposite)
Felling a spruce tree in a Devon woodland, using a power-saw.

The wood from the trees

The wood from the trees

How a circular saw cuts a round log to a squared outline. The bench on which the log rests is moved steadily forward towards the whirling, sharp-toothed blade.

Sawing and smoothing timber

Take a look at saws, and note how their two ranks of teeth cut out sawdust. This basic principle is used in all sorts of saw patterns. Think of big cross-cut felling saws used by tree-fellers who work in pairs, small 'one-man' bow saws, rip-saws for carpentry and joinery and great circular saws and band saws in a saw mill. Nowadays most trees are felled with a motor-driven chain saw, a fascinating although noisy machine that needs training and skill to handle. See if you can find a feller at work in the woods.

At a sawmill the main work is the cutting of logs along their length into planks of various widths and thicknesses. Other saws, with different teeth, make the cross-cuts. You are not likely to see much of all this because the use of high-speed power saws makes a mill a dangerous place to visit. But you can easily see what goes in and what comes out. Obviously much material is lost when round logs are cut into

The wood from the trees

A wood-turner shapes a bowl; the well-seasoned piece of walnut is spun in the lathe against his sharp chisel, which cuts out the round hollow required.

square-edged planks. Can you calculate *how much* wood is lost in cutting one likely size of log into planks? Saw millers aim to use waste wood for fuel or chipboard, but they burn most of the sawdust, for they make more than anybody can use. Look for the long tubes through which sawdust is sucked from the saws to the burner.

All saws leave rough surfaces, because of their two rows of teeth. Smooth surfaces are secured by using the carpenter's plane, or else by high-speed planing machines with rotating blades. All wood you find with a smooth surface, whether it is part of a building or a piece of furniture, has been planed.

A tropical rain forest with tall broad-leaved trees rising from buttressed roots on swampy ground, slender palms and trailing climbers.

Carving, turning and veneering

Other processes that are applied to high quality wooden objects demand special skills and tools. Wood carvers use chisels and mallets to cut intricate designs on wood surfaces, or deep into the substance. They use well-seasoned selected material, held firmly in a vice. Turners use a lathe that 'spins' a piece of wood round rapidly as they apply chisels to shape it. They can only make round articles, but there are many forms of these, including chair legs, lamp stands and fruit bowls. See if you can find one of these people in action, or

The wood from the trees

else study examples of their work at home, in shops or in a museum.

Veneering means the application of a thin sheet of a costly, decorative wood to a base made of another kind. A walnut wardrobe will have walnut only on its surface, its main structure being, perhaps, beech.

The earliest veneers were sawn by hand but modern veneers are cut from selected logs which are first softened by steam and then either forced or rotated against sharp peeling-knives. Veneers are so thin that it is not easy to tell where they have been applied.

Transporting, seasoning and preserving wood

In the past the work of felling, shaping and smoothing timber was done by country craftsmen, and trees were converted into houses and furniture within a few miles of the spot where they grew. Nowadays most stages of the process are mechanised, and are carried out on a huge scale, often involving transport over long distances, possibly half-way round the world.

Have a look at the many ways in which round logs, sawn timber, and wooden manufactured goods are moved. In the forests, horses, oxen, elephants, buffaloes and even camels have been used in different countries.

Nowadays the usual methods require a tough tractor which can load logs on a trailer and move across rough ground, or else operate an overhead cable crane. Specially-designed lorries, called 'pole waggons', are used on highways. Timber also travels overland by rail, and some remarkable forest railways have been built. In some countries logs are floated down rivers, or made into rafts that are towed across lakes. Look for examples and pictures of these varied transport methods.

When trees are first felled, even in winter, their trunks are saturated with water — remember that they carry sap. If people use wood in dry situations, this water evaporates, making the wood much lighter and slightly stronger. But it also shrinks unevenly, and if carpenters, joiners or furniture makers were to use unseasoned wood the things they make would get smaller to different degrees in different directions;

they would warp, and cease to fit. The timber merchant anticipates this by 'seasoning' — that is drying — wood in advance. Look out for high-grade thick oak planks slowly air-seasoning in a timber yard. Timber for everyday use in bulk — in house-building for example, is artificially seasoned in heated kilns. See what you can find out about this process and say why it costs less than air seasoning.

Everyone knows that if logs are left on the damp forest floor they eventually rot away. They decay because they form the food of certain fungi and insects. People often use wood in similar wet places, for example as fence posts, but they want it to last a long time. The only way to be sure of this is to apply a preservative, either a tar oil like creosote or a poisonous metallic salt. This sounds simple but special processes and plant are needed to get the chemicals deep into the wood.

Find out more about one possible process.

Above ground level, outdoor woodwork should always be protected by paint or varnish. Ask a painter or handyman how he does this; why are two or three coats needed?

Fine ornamental timbers, used for furniture indoors, are never painted. Instead the beauty of their natural grain is revealed by skilful polishing to work wax into their surface and get a smooth finish. Sometimes they are stained to a different colour, usually a darker one, and this can make timber identification tricky. Without studying too much detail, you should learn to recognise these processes, which add a great deal to the natural beauty of decorative woods.

'Man-made' wood, board and paper

Wood in its natural solid form is fairly easy to study, all the way from the growing tree to the final manufactured article. But about a third of all timber harvested today is reconstituted or 'man-made' into some form that makes its original character hard to recognise. This is done in factories, by mechanical and chemical processes difficult to follow. Each final product really calls for a separate inquiry, so unless you have some special interest, it is best to make only a simple outline of what goes on. These materials are very important commercially and give foresters a large part of their income.

1. *Plywood* is the simplest form of man-made timber. Selected logs are first softened by steam, and then rotated against a sharp knife in a huge, powered lathe. This peels off long thin sheets of veneer from the surface of the log. Three-ply wood is made by glueing together, very strongly, three layers of veneer, rather like a sandwich, with the grain of the outside pieces going one way and the grain of the inside piece going across the other way, at right angles, so that the board keeps flat and is hard to split. Thicker boards can be made by using more plies — see if you can get hold of a piece of five-ply, seven-ply or even nine-ply wood, and make a note of what they are used for.

2. *Chipboard* is made by cutting logs or waste lumber into chips, then sticking them together into large flat sheets with a plastic resin, under heat and pressure. It uses up cheap wood, and provides big flat panels of any convenient size and thickness, for building, shopfitting and furniture manufacture.

3. *Hardboard* consists of thinner, harder sheets. Insulation board is softer and thicker and is used to hold in heat or deaden sound. Both are made by shredding cheap timber into fine fibres, then compressing it after adding plastics to make the fibres stick together.

Handymen often use all these materials, so samples are easily secured. Make and label and measure a set.

Paper, and paper-like materials such as cardboard and art boards, are nowadays nearly all made from wood pulp. Logs, mainly softwoods, are ground down into tiny fibres by grindstones working under water, or else they are broken into chips which are cooked in strong chemicals until their fibres separate. These fibres are then floated, in a vastly greater volume of water, on to a wire mesh screen that lets the water drain through.

The wet fibres immediately mesh together, and as the pulp is quickly dried they automatically form a sheet of strong paper. All this is done as a continuous process on a huge paper-making machine, working at high speed. Wet pulp flows in at one end and rolls of paper come out at the other. There are many variations that give all the different thicknesses, textures, and colours of paper we need, with different surfaces. You can easily make your own collection to show the great variety of this 'reconstituted' wood product, and its many uses.

Other materials obtained from trees
Can you think of materials, besides wood and paper, that men get from forest trees? Examples are syrup and sugar obtained by tapping Canadian maple trees in early spring, resin won by cutting into the trunks of pine trees in south-west France, and the rubber that oozes from cuts made in the thin bark of the rubber tree grown in Malaysian plantations. Then there is cork stripped as bark from the cork oak tree in southern Portugal, and tanning materials obtained, for example, from oak bark in England or wattle bark in South Africa. You may decide to follow up the full story of one of these substances. List as many tree-based materials as you can.

Many uses for timber
Now that you have discovered how wood is converted to the many forms in which we use it, make a search for good examples and list them. Nearly all old houses and farm buildings in England are framed in native oak, and so are church roofs. Look for big tithe barns, 'cruck' cottages, and the 'king-post' or 'hammer-beam' structures in churches. Choir stalls, pews, altars and other church furnishings will usually be oak. A book on the history of architecture will help you to interpret and appreciate all this.
If you like ships and harbours you can see wood used in many ways, including wharves and lock gates. You may see a cargo of softwood from Scandinavia being landed. In Scotland and northern England sturdy fishing boats are still built with oak frames and larch planking, and you may come across a shipyard where they are being made. Transport and communications use wood in many ways — think of telegraph poles, railway sleepers, packing cases, and pallets or wooden platforms on which many sorts of goods are stacked

The wood from the trees 59

Building a wooden fishing boat at Arbroath in east Scotland. The keel and ribs are made of oak, while the planking will be made of larch, all assembled by hand.

for movement and storage. Then there are engineering works and mining. Although steel and concrete do most of the load-carrying today, wood is used during the construction stages of bridges and docks, and for pit-props. On the farms the main use of wood is in fencing but you will also find barns, cow-stalls and implements still needing timber. Try listing all the different ways of employing wood that you can see in a short outdoor walk.

Indoors the scope is endless. Select a building, such as a house, school or workshop and note how many things have been fashioned from wood. Looking around me now I see chairs, tables, lamp-stands, cake-stands, picture frames, a fender, trays and a grandfather clock. To pursue this line further would involve a long study of the history of furniture. I can go into the kitchen and find a rolling-pin, bread board, spoons, bowls, and handles for cooking knives, sieves and ladles all made from wood. In the toolshed outside stand spades, trowels, forks, rakes, a saw and an axe, each with its wooden handle, ready for garden use. When you come to

Making a cricket bat. The craftsman is tapping the handle, made of split bamboo cane, into a blade that has been skilfully shaped from a hand-cleft piece of willow log.

describe things like these, don't forget your sketches. Advertisements can provide many illustrations, especially ones of modern objects on sale today.

Descriptions of different timbers in chapter 3 will suggest what kinds of trees have provided many of these wooden objects. For your general account of wood in use, a provisional identification — such as 'probably oak', 'possibly mahogany' or even 'ash(?)' will suffice. Only an expert can identify timbers certainly from surface inspection. Do not spend too long on naming woods, but note how each object has been made. Has it been turned, carved, veneered, stained or polished? Was it mass-produced in a factory or hand-made as an individual creation in a craftsman's workshop? Look at a wide range of wooden goods, for only in that way can you appreciate the remarkable versatility of timber taken from the living tree and shaped to man's needs.

Forests throughout the world 7

Studying a forest region without going there yourself
Other forest regions
Forests in the future

In this, the final stage of your project, you should take a look at the history and geography of a large forest region, and think out what its future is likely to be. Select an area that appeals to you, such as a large slice of a continent, a single country, or possibly one forest like the Forest of Dean in Gloucestershire. You cannot consider this forest region in isolation from the rest of the world, for as we saw in the last chapter, the timber trade is world-wide. You will have to bring in other aspects of the region's life and trade, its transport system and ways in which people use its other land, such as for grazing sheep or growing crops.

How to study a forest region
You will be dealing with a large expanse of country, and with what happens there over a long spell of time. If it is a distant land that you cannot visit, you will have to rely on other people's reports in books and journals, or possibly in trade advertisements or tourist pamphlets. Few books describe particular forests, and even these are often out of date. Look for a chapter on forests in a general account of a country. Explore your school library first, then see if you need help from other librarians. See if you can supplement book work with closer contacts. You may perhaps visit a forest in Britain or Europe, meet a traveller who knows the Canadian backwoods or the African jungles, or collect samples of timbers, tree leaves or fruits that illustrate the life of distant woods. Nearly all the trees that grow in temperate-zone forests can

Forests throughout the world

be found and seen alive in arboreta — that is, collections of rare trees — somewhere in the British Isles. A trip to a zoo will bring you face-to-face with many jungle beasts and birds — safely behind bars! Failing live animals and birds, look at a collection of stuffed ones, possibly in a large central museum, such as the British Museum's Natural History building at South Kensington in London. See what specimens were shot in your chosen forests. Local museums, too, may exhibit similar selections. Other groups of specimens linked to forests are plants, fruits, timbers, modern manufactured wooden goods, and old-fashioned wooden tools and household implements, often referred to as 'bygones.' The ethnography section of a big museum, like that of the British Museum near Piccadilly (Burlington Gardens) displays the tools, household goods, weapons and boats of many forest peoples. For example you may find drinking vessels and spear shafts made of bamboo, a woody grass that you seldom see in temperate lands.

Forests at the tree-line of the Alps above Arosa in Switzerland. These natural woods of pine, spruce and larch check soil erosion and stop avalanches sweeping into the valleys; they are never clear felled.

Forests throughout the world

If you are studying languages, you may find books in, say, French or German describing the forests and woodland customs of those countries. If they are illustrated you can quickly extend your knowledge of both the language, the people, and the trees.

An example of a detailed study

To see what sort of facts you should collect, let us look at the forests of British Columbia, the westernmost province of Canada, which borders the Pacific Ocean. From your atlas and a geography textbook you can discover the size of this region in square kilometres, define its situation on the earth's surface by latitude and longitude, and see its relation to the North American continent. A physical map of North America shows that it is mountainous, with frequent hills and valleys. A climatic map will disclose its high rainfall, giving rise to large, swift rivers. A map of vegetation will tell you that much of it is under coniferous forest.

Discovering which trees grow there is not so easy. You will need to find books on the geography and trade of Canada to find out that leading conifers in the west include Sitka spruce, Douglas fir, Western red cedar, and Western hemlock, all unfamiliar names. It so happens that all these grow well in Britain, so you can look for living trees in parks, woods, or Forestry Commission plantations, observe and sketch them at first hand, and collect specimens.

Now follow up the trade in their timbers. Figures for area, trade and population will reveal that this big territory holds relatively few people but sends timber, and timber products such as paper, all over the world. But this is a fairly modern development. Work out why this great forest resource was neglected until the present century. What difference did the opening of the Panama Canal make?

How is the timber felled and transported to the seaports? If you look in books published many years ago you may find remarkable pictures of tools, methods and machines now considered out-of-date. The lumbermen will be swinging huge axes or pulling on great two-man cross-cut saws, and horses or ox-teams may be hauling out logs. Daring engineers

Forests throughout the world

Canada's Rocky Mountain forests, set between ice-bound peaks and still waters, provide magnificent scenery as well as timber. Silver firs beside Lake O'Hara in the Yoho National Park, British Columbia.

may be shown rigging overhead cables to tall trees, so that steam engines can draw logs to quaint railways, where other steam engines, with big 'spark-catcher' chimneys, will haul heavily-laden trucks to the ports, crossing high lattice-work wooden bridges on the way. More recent illustrations will show power saws and huge mobile cranes that haul in the logs and then load them on to enormous lorries. Every man on the job will be wearing a bright yellow safety helmet, unknown in the days of the pioneer backwoodsmen. Do not discard the facts you gain from the older books and pictures. Change in the lumber industry is all part of the story, and your studies should reveal it.

Water transport has changed less dramatically. Logs are still floated down rivers as before, built up into great rafts when they reach the sea, and then towed down the coast by powerful tugboats.

Of course, all this activity is changing the forest itself. Great

Forests throughout the world

gaps are being cut in the original forests, but the timber firms want to come back some day to harvest another crop. Find out what is being done to re-stock the cleared land. Will the natural seed that falls from surrounding trees replenish the crop, or is planting done too? Has the province got a sound forest policy to preserve its timber reserves? If so, who sees that the right things are done, and who decides how fast the woods may be felled? Think of the big loss to the country if timber was cut too fast. Saw-mills and paper-mills would be forced to close down, and valuable exports would cease. Worse still, the steep hillsides would suffer erosion, losing their topsoil which — in the absence of forest cover — would be carried downhill as silt to block rivers and harbours.

Now take a look at the wild life of these Canadian forests. The plants, birds and insects are much like our own, though they belong to different local species. Detailed information on all this can only be found in specialised books, so it is simpler to concentrate on the mammals, that is, the four-footed beasts. Some are closely related to kinds that once roamed wild in Britain but are now extinct. Think of the wolf, the bear and the lynx. The deer are similar to ours but have different names and specific characters. See how much you can discover from books written by naturalists and huntsmen who have lived in British Columbia or in neighbouring lands.

Lastly consider the people and their history. When did the first European explorers arrive and what races did they find already established? One of the main Red Indian tribes was called the Haida. Because they lived in forests along rivers and the sea coast their way of life differed from that of the Indians on the prairies. For example, they had sound wooden houses instead of portable wigwams, and were skilled sea-farers and fishermen. They carved canoes out of the light, strong, and lasting wood of the Western red cedar tree (see chapter 3). They used the same timber for their totem poles, and you may find a real example in some large museum. Find out all you can about totems, those strange monuments designed to record the history and ancestry of each tribe. Take a look, too, at the story of immigration and the rise of modern communities. What can you discover about

Vancouver, the capital of the province and a great seaport dependent on the forests for its export trade?
If you can find an official report by the provincial forest service you may discover how much woodland is owned by the government, by lumber firms and by private individuals. It may also say how much timber is cut each year and how much land is replanted. But do not spend too much time hunting out statistics. A few main ones, such as the number of men at work in the woods, will suffice; details can easily confuse the picture.
What perils threaten the forest? Fire is the worst, and you may discover what areas of woodland have been burnt recently, and how the fire-fighters are trained and equipped. They are sure to have look-out towers and fire-spotting planes, and some may even jump from aircraft with parachutes. Insect pests and great windstorms may have done damage, too. Look at your forests from all possible angles, considering losses as well as benefits.

Other forest regions
There are many other great forest regions that may appeal to you as good subjects to study, possibly because you already know the country or its language. In England there are several ancient royal forests, usually with oak as their leading tree. These include the New Forest in Hampshire, Gloucestershire's Forest of Dean, Sherwood Forest near Nottingham, and the old Wealden Forest of Kent, Surrey and Sussex. Local historical books and papers can shed a lot of light on how these forests were first preserved by Norman kings for hunting, then slowly cleared for pasturing flocks and herds, and cultivating corn crops. Their timber was used for building houses and ships, and to make charcoal for medieval iron, glass, pottery industries.
Since 1919, there has been a big national effort to restore the woodlands of England. This has been led by the Forestry Commission and helped by private landowners and conservation bodies, including the Countryside Commission, the National Trust and the Nature Conservancy. Much of today's replanting is, however, done with coniferous trees,

Forests throughout the world

Using a tractor-powered cable-crane to haul spruce logs to a forest road.

whose softwood timbers grow faster, thrive on poorer soils and are in greater demand by industry.

In Scotland, well-known historic forests are few, but there is a vigorous drive to replace the vanishing natural pinewoods with productive forests of larch, spruce, pines once again, and Douglas fir.

In Wales, most of the ancient forests became treeless sheep pastures, but since 1919, when the Forestry Commission was set up, nearly one-tenth of the land has been planted with coniferous trees. Can you think of reasons why so many outlying sheep farms no longer provide a good living for their owners?

Looking overseas, you could study Alpine forests in Switzerland and neighbouring countries. These have trees like ours, but many grow far up mountains which are much higher. Many Swiss forests protect upland pastures and villages from avalanches; therefore they must never be 'clear felled' for whole valleys would then become desolate.

Work out how trees have this safeguarding effect. How can such 'protection forests' be worked for timber? What is their value for scenery and recreation in a country with a big tourist trade?

Or you might choose some tropical country, such as a province of India with completely different jungle trees. These may include timbers like teak that are marketed and used all over the world. Here you will find another range of wild life, including tigers, elephants, monkeys, parrots and pythons. The forest dwellers will also be strange, with unfamiliar religions and customs. Their buildings will be made from unfamiliar local forest materials, including bamboos for walls and palm-leaf thatch to make a dry roof. Remember that the climate is very different, too, with high temperatures, no frost or snow, perhaps a dry season and a very wet one, or maybe rainstorms all the year round. You will need to go into the geography of your chosen country thoroughly before you can understand its forests and the part they play in its economy.

Forests in the future

Lastly, round off your study by taking a brief look at forests in relation to man and the world as a whole. Trees occupy about a third of all the land there is, though the distribution of woodland is very uneven. Consider how valuable the forests are for protecting the soil by checking its erosion by water and weather, how they help to regulate the flow of streams, and give shelter to crops, livestock, and buildings. Why is timber such a valuable item of trade, constantly being sold by countries with a surplus to those with a shortage? Do you think that a *world* timber shortage is possible, and if it

Forests throughout the world

happened, how would people suffer?

Experts in the Forestry Division of the Food and Agriculture Organisation of the United Nations are constantly collecting figures and encouraging governments to make plans to keep forests growing for all foreseeable time. But people will only support this action if they are convinced that thriving forests are essential to their own well-being. If your project has been thoroughly done, it will reveal why this is so, besides having introduced you to timber's wide range of uses and to the fascinating life of the woods.

Measuring trees

Height
Start with a first estimate. Judge the tree's height by that of a man, a lamp post or a neighbouring building. Improve on this by making your own measuring rod. Take a long stick, mark off two metres along it, with the aid of a ruler, and cut it across at that point. Use this rod to measure the lowest two metres of the tree, from the ground up. Then hold it upright beside the tree and get a friend to estimate the remaining length of the tree trunk. He can do this by standing well back, covering the image of the stick with a pencil held up in front of one eye, and seeing how many pencil-lengths (each equal to two metres) he needs to reach the top of the tree. This is only one of many ways to ascertain the height of a distant object. A shadow method is described in chapter 3 (page 18). Can you think of others? If so, try one out and say how you did it. Compare the results from two methods.

Thickness and cross-sectional area
How thick is the trunk? Put a tape measure round its base and measure the circumference in centimetres.
Now find its diameter. To do this divide the circumference by the mathematical factor called π which you can take as $\frac{22}{7}$ or 3·142.
Next, work out the area of the cross-section of the base. To do this, divide the diameter by 2, to get the radius. Then apply the formula: $\pi \times \text{radius}^2$, to get the area in square centimetres.

Volume
For sale purposes, timber is measured by volume. *If* your trunk were cylindrical, you could find how much timber it held by multiplying its height in metres by the cross-sectional area in square centimetres. Because there are 10,000 square centimetres in one square metre, you must divide the result

Measuring trees

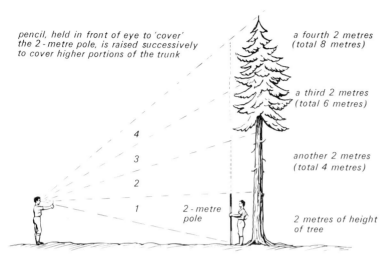

Measuring the height of a tree.

pencil, held in front of eye to 'cover' the 2-metre pole, is raised successively to cover higher portions of the trunk

a fourth 2 metres (total 8 metres)

a third 2 metres (total 6 metres)

another 2 metres (total 4 metres)

2-metre pole

2 metres of height of tree

by 10,000 to find the volume in *cubic metres* — the unit that foresters use. This is a 'big' unit, so do not be surprised to find that this trial answer is only a small fraction of one cubic metre.

But your trunk is *not* a cylinder; it tapers towards the tip like a geometrical cone. Remember the formula for a cone's volume; *one-third* base x height. Then divide your trial answer by 3 to get the correct one.

Example (with simplified figures)

A larch tree is 6 metres tall and 25·1 centimetres round at its base. Divide 25·1 by the constant π, i.e. 3·142 (or by $\frac{22}{7}$), to get the *diameter*: 8 centimetres. Halve this to get the *radius*: 4 centimetres. Square this: 4 x 4 = 16, and multiply the result by the constant π i.e. 3·142, to get the *cross-sectional area*: 50 square centimetres.

Multiply this area by the length (6 metres above) and divide by 10,000 to get the *volume*, in cubic metres, of an equivalent *cylinder*: 6 x 50 = 300 .˙. volume = 0·03 cubic metre.

Divide by 3 to get the volume of an equivalent *cone* (that is, your tree trunk). This is: 0·01 cubic metre.

How many trees this size do you need to get one cubic metre of timber?

Foresters apply these simple mathematical formulae daily to discover how much timber they have in their forests, using

Measuring the timber volume of a felled oak log. Using a tape measure, the foresters take the length in metres. The circumference is then measured half-way along, to obtain an average area for the log's cross-section, calculated in square centimetres. Further sums then give its volume, in cubic metres.

tables to get quick answers. In practice, they usually take the tree trunk's circumference at 1.3 metres above ground, a convenient height for measurement, and make a final division by 2 instead of by 3, a 'rule-of-thumb' that works well. Once a tree has been felled, it is easiest to measure the circumference half way along each log. (See the photograph above.) This gives, in practice, an acceptable average cross-section figure for volume calculation, without any division or adjustment.

Naming trees

Besides its common English name, each sort of tree has a scientific name, in Latin, which acts as an 'extra' label to make its identity more certain. There are four parts to this label. Scots pine for example, is called: *Pinus sylvestris* Linnaeus (Pinaceae).

The most important part of the name is the first word *Pinus,* usually printed in italics with a capital letter. This gives the *genus,* and all trees in the same genus are much alike; generic characters are the ones to study first. The second word, *sylvestris,* in italics but with a small first letter, tells you the *species.* The specific characters that distinguish Scots pine from Lodgepole pine, *Pinus contorta*, are small, but matter a great deal to the forester who may plant either tree by the thousand. Lodgepole pine differs from Scots pine (see page 29) in having deep-green needles, tiny prickles on its cone scales, and black bark.

The last two words in the label are only important to serious botanists. 'Linnaeus' is the name of the man — a famous Swedish naturalist — who first described this tree carefully and published this scientific name. '(Pinaceae)', usually in brackets, records its natural family, but only an expert can explain why it is classified in this 'pine family' rather than in another similar one.

Scientific names are very useful because they are international. A Frenchman calls the Scots pine *sapin rouge*, and a German calls it *Rotkiefer*, while many timber merchants call its timber 'redwood'. But all agree that it is *Pinus sylvestris* L., the letter 'L.' being the common abbreviation for 'Linnaeus'.

Handling specimens

Leafy shoots
A leafy shoot, showing two or three leaves, some buds, and the character of the twig, is the best unit for record and identification. It shows how the leaves are placed along the shoot — in opposite pairs in some kinds, singly in others. When picking leafy shoots, take them some distance back from the leading shoot or main branch of the tree. This avoids harm to its growth, and gives a more typical specimen. Leafy shoots of broad-leaved trees are easily preserved by pressing them between sheets of newspaper under moderate weight, such as a heavy book. When dry, they can be mounted on stiff paper, using transparent sticky tape, and labelled. Many conifers, however, shed their needles when they are dry. So to be sure of keeping a record, make sketches as well, while everything is still fresh. A handy method is to fill an exercise book with sketches *and* pressed specimens, all named.

Flowers and catkins
These can be pressed also. It helps identification if you can gather them still attached to a shoot. You can only gather them at certain times of year, so keep a note of collection dates.

Fruits, cones, seeds and bark specimens
These are best kept in small transparent plastic bags, with labels attached. Juicy fruits will dry and shrink, so sketch them while they are still fresh.

Wood
Look out for small logs and bits and pieces of wood in a

Handling specimens

partly manufactured form. The waste or 'offcuts' from a joiner's bench or furniture maker's workshop can tell you a good deal about the character of a timber, especially if its identity is known. Aim to get two pieces of each kind, so that you can keep one intact and cut through another to reveal its grain.

General
Be sure to label and date everything, and to keep a separate record of all that goes into your collection. Put down the place where you found it, too, as such important details are soon forgotten.

Aim to link each aspect of your collection with the main project. For example, if you write about 'Trees in Hollybush Wood', your essay, sketches and photographs will become much more real if a leafy shoot from each kind of tree, and a piece of wood of the same sort, is added to round off your story.

Books to read

General accounts of forestry
Herbert Edlin has written:
Trees, Woods and Man, Collins, New Naturalist.
What Wood is That? Thames & Hudson.
Guide to Tree Planting and Cultivation, Collins.

Other helpful, well-illustrated textbooks are:
The Seasons and the Woodman by D.H. Chapman, Cambridge University Press.
In a Wood by M.M. Hutchinson, Ward Lock.
A Wealth of Trees by A.C. Jenkins, Methuen.
Let's Look at Forestry by I. Lewer, Warne.
Woodland Ecology by E.G. Neal, Heinemann.
Woodlands by J.D. Ovington, English Universities Press.
Teach Yourself Forestry by T.A. Robbie, English Universities Press.
Trees and Forestry by G.E. Simmons, Longmans, Green.
Timber, The Story of our Forests by B. Taylor, University of London Press.
The Scots Pine – An Introduction to Forestry by H. Watson, Oliver & Boyd.
Thanks to Trees by I.E. Webber, World's Work.
Science and the Forester by L. Wolff, Beal.
Growing and Studying Trees by J.B. Wood, Blandford.

Identification of trees
Herbert Edlin has written:
Wayside & Woodland Trees, Warne.
Treasury of Trees, Countrygoer Press, Buxton.
Know Your Conifers, Forestry Commission Booklet 15, HMSO.
Know Your Broadleaves, Forestry Commission Booklet 20, HMSO.